Walter Gith

Love and Physics

Publisher: BoD · Books on Demand GmbH, Überseering 33,
22297 Hamburg, bod@bod.de
Print: Libri Plureos GmbH, Friedensallee 273, 22763 Hamburg

ISBN: 978-3-7583-4264-6

Love and Physics

Love and Physics

Consciousness: A self-referencing mechanism

Walter Gith

Love and Physics

LOVE AND PHYSICS

Content

Preface

It has been almost 35 years since the German edition of this book has been published. Every topic in the book is still valid as it was written so long ago. The reason is that there is no new knowledge presented. I just did a rephrasing of ages old knowledge, in order to make it palatable to the modern reader.

I postponed the translation over and over again, because I was too lazy. Some 10 years ago I tried Google translate and it was a disaster. When I recently met someone very interested in reading my book in English, we just tried Google translate again. This time it was a major success. Hence, I can present the English edition to you here.

I have to point out, that the spiritual and religious references are looked at as examples. In no way was any reference to the spiritual beyond or any spirit in the beyond at all intended. The examples were just taken out of our cultural background and are therefore ingrained in the sub-consciousness of so many people.

Also, when I talk about *God*, I do not mean any spirit or being in the beyond. On the contrary, for me *God* is very real and represents everything there is. Whether the term *God* should be used representing the whole Universe is debatable, especially in the light of latest discoveries of Physics. In Physics it turns out that the Universe is possibly not everything there is!

Sincerely
Walter Gith

Introduction

We all have great respect for the scientific method. It has given us great comforts. We can pick up the phone and talk to a friend in a faraway country. Or we can get on a plane and just fly there. We use electricity to make light for ourselves in the evenings, put the used dishes in a dishwasher and then just clean them out again. The convenience of scientific progress has not only led us to admire the scientific method and appreciate its results, but we have already acquired its underlying logic. Not only are we believers in science, but we are all scientists without knowing it. However, science has developed faster than we ourselves. Science has already completely changed its old views and advanced to completely new ideas about the world. Unfortunately, these ideas are only discussed within small groups of specialists and therefore do not reach the general consciousness.

Therefore, with this book I try to redefine some apparently familiar terms or put them in new contexts. In doing so, I hope to reduce problems of understanding that exist today between the different areas of life and research. The new connections may not only promote your own understanding, but also the understanding between people.

Because it is not science as a research body that determines world events, but rather influences of each individual determine events in the world.

Our world today therefore reflects the state of our own development. Let us try to learn from modern science and especially from physics as a pioneer.

The first step is to become aware of your own outdated scientific way of thinking. By recognizing one's own scientific methodology, it becomes easier to join the new path of scientific development and thereby steer world events in a positive direction.

Therefore, in this book we will not only apply the scientific methodology, but also disclose it so that everyone can follow it to the fullest extent.

The decisive result of the new research is that our so-called external reality is not independent of our own consciousness. The nature of reality is directly dependent on our conception of this reality. Our perspective and our models directly determine our physical reality. If we change our ideas about reality, reality itself changes. Improving our thoughts therefore leads to improving our reality.

The first four chapters of the book create the conceptual foundations of a new integrative thinking that leads to better understanding and can thereby improve our reality. The 5th chapter summarizes the new connections again in a compressed form. This form is suitable for reading in meditation or deep relaxation. Our minds are very sluggish and have very limited receptivity to new things. It is therefore advisable to anchor the new thoughts in very deep layers of consciousness so that they are constantly present to us. Only constant presence leads to a lasting change in reality. Intellectual knowledge is not enough; it must be tested and deepened in practice.

The introduction of the concept of God is particularly important to me. The concept of God that appears here is related to that of the known religions, but differs significantly in its derivation and scientific understanding.

The profound certainty of our existence and therefore of God (=all that is), both from theoretical considerations and through an intuitive understanding through inner perception in meditation, is the crucial message in this book. One can overcome belief in God and achieve an awareness of God in which the presence of God is felt!

Once we have recognized and experienced that our thoughts shape reality, it is only logical that we integrate something divine into our thoughts and ideas. Because only the assumption of something absolutely good or absolutely harmonious ultimately leads to this state.

I am aware that the knowledge presented in this book has been known for thousands of years. Of more crucial importance, however, is the constant reformulation of this knowledge in order to establish a connection to today's understanding using appropriate language. Quotations should therefore be seen less as *evidence* of what is already known, but rather as an alternative formulation to the one written down here.

1   Communication and information

The communication model

To show what is meant by the "scientific approach," we will select a process from our daily lives that everyone knows and practices all the time. We examine the process of communication. In every communication, information is exchanged without everyone being aware of the exact mechanism of this information exchange.

As a special form of communication, we examine perception and here again the exchange of information with the environment as the smallest communication unit.

We begin in the tradition of scientific approach by observing and describing the act of communication. We first use the terms in their general meaning and then later arrive at more precise definitions of the terms.

Imagine, dear reader, that you are walking through the jungle in Brazil with a knowledgeable native. He shows you different animals and plants that you have never seen before. For a particular plant, he points out that this plant is edible and lets you taste it. He recognizes that you like the plant and tells you the name of the plant, "Haldala". Haldala is very common in this area and can therefore also be bought in smaller villages. The native probably told you the name so that you could ask about the plant if necessary.

What happened? Apparently, an exchange of information took place. You have received information about the appearance, edibility and name of a plant. If the whole thing was more of an exciting experience for you, you probably took in a lot more information.

Now the scientific investigation of the process begins. We first put together the channels through which information - transmission is possible. Obviously, the five senses are transmission channels because they are based on physical carriers. 1. Vision, this is where photons or light waves are transmitted. 2. Hearing, where sound waves are transmitted through the air. 3. Touching, here direct contact of matter leads to transmission. 4. Tasting, here too direct material contacts lead to transmission. 5. Smelling, here the smallest particles of matter (molecules) are transferred.

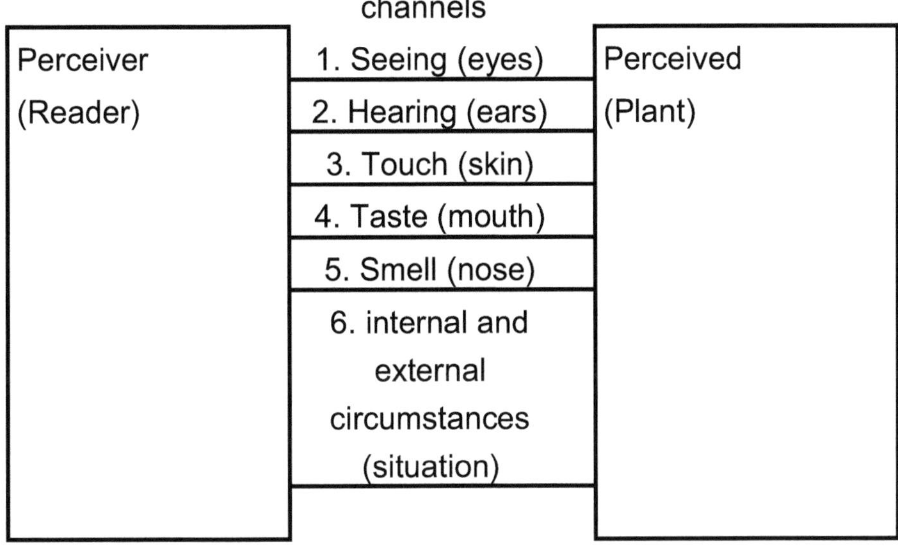

Figure 1.1: Perception model

But there is also a 6th channel that has no carrier. It is a logical channel, but it does influence the information transmitted. The following consideration should show how much the information content that you received about the plant during your jungle walk depends on the environment (i.e. on the sixth channel):

Imagine that, in order to make the walk a special experience for you, the native has chosen a plant that is not normally used as food, but is used to have a slightly euphoric effect. Because of the plant's delicious taste, you assume that this plant is intended for normal consumption. In your general excitement about the jungle walk, you don't consciously notice the plant's euphoric effect.

Due to your personal life situation, you now initially consider the plant to be a food, while for the natives the plant is clearly a pleasure enhancing item. The different life situations (channel 6) lead to very different information about the plant. Your ideas regarding Haldala are very different from that of the natives.

The introduction to our topic was taken from real life and is therefore of a complex nature. We could perhaps fill several books if we wanted to describe the visit to the jungle in detail, that is, taking all channels of perception into account. Since we want to get to know the scientific methodology, we will initially limit ourselves to the appearance (visual image, 1st channel) and the name (verbal image, 2nd channel) of the plant. The aim is simply to examine the connection between a specific plant and a specific term.

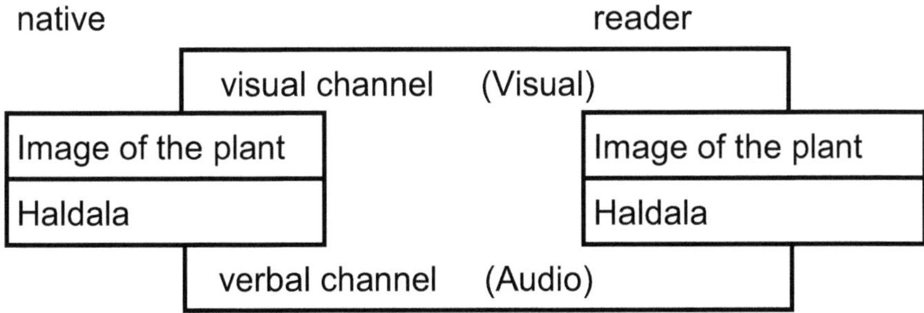

Fig.1.2: Creation of identical meaningful contexts/meaningful spaces (MS)

In a shared living situation, it is easy to get a common concept of a common thing: the plant is seen and named.

The identification of the plant with the term Haldala creates the prerequisites for a new communication. Both communication - partners have now developed the same concept of the same thing; they have formed a context of meaning. In this way, any number of meaningful connections can be formed. *We want to call a number of those connections 'meaning space'.*

What is characteristic of the above synchronous structure of a meaning space is that there are at least two communication channels. In our example, on the one hand, the visual channel through which the image of the object or item is conveyed, and on the other hand, the verbal channel through which the name is pronounced and heard.

In order to arrive at a simple, scientific model, we now want to restrict our view of reality even further by looking at the special case of single-channel communication. To do this, we make the assumption that identical spaces of meaning have already been defined. So everyone involved knows which plant "Haldala" is. Using the verbal channel, the corresponding object can be created in the recipient's mind by saying the name. In our example, it is enough to say the name "Haldala" and everyone involved knows which plant is meant.

After making enough restrictions and idealizations, we can take the scientifically significant step and move to a general model of communication.

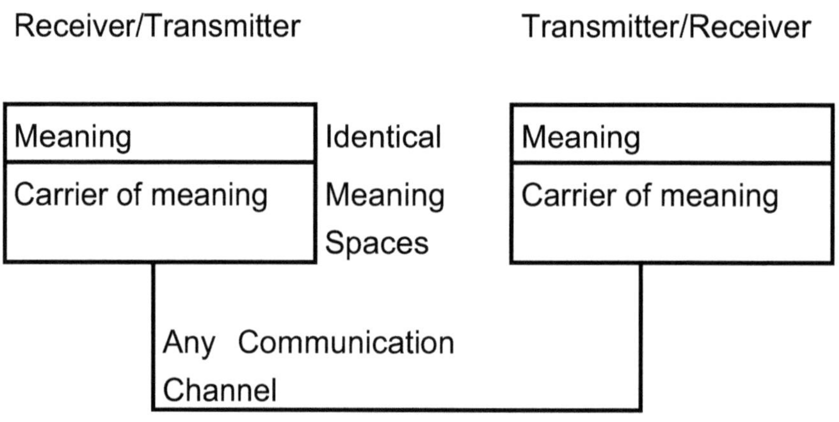

Figure 1.3: Communication model

In fact, making restrictions and idealizations when finding scientific models is a characteristic approach. We summarize the so-called boundary conditions again:

- We assume that identical meaning spaces already exist.
- Communication can therefore be reduced to one channel.
- The communication channel is no longer firmly defined in the model, but can be of any type.

The above model is actually the most common communication model today. It is mainly used in the technical area. Despite all its limitations, idealization has the enormous advantage that the resulting model is easy to communicate and therefore understandable by everyone.

This approach has contributed to the widespread dissemination and thus the success of science. Anyone can use such a model at any time and check its accuracy. However, it is important that the idealization undertaken, i.e. the limitations or inaccuracy of the model, always remains conscious. This is exactly what has not always happened in the history of science. As a result, a false understanding of science spread. The sometimes simple models/ideas of the world (e.g. mechanistic world view) have spread rapidly without their limitations and idealizations being communicated. The scientists who did not take this into account are certainly partly responsible here.

In the following, we will get to know how models work by applying our model to our small example: The communication channel is language, i.e. the verbal channel. The word "Haldala" carries the meaning. The assigned meaning is the plant. If you, dear reader, want to buy the plant in question after your walk in the jungle, this is exactly how our new model comes into play:

you go to a local shop, ask for Haldala and (hopefully) get the plant you want!

A closer look shows that the purchase of the plant can only be represented by the simple communication model under very specific circumstances.

For example, it may be that your meaning space and that of the salesman are not identical in the crucial point, so that the natives in your shop laugh loudly or grin when you ask for Haldala. This immediately expands your own scope of meaning because you now know that there must be something special about the purchase or the plant. You'll probably try to figure out why people laugh. You will probably learn about the special effects of the plant and you will have expanded your scope of meaning again. People may not want to sell you the plant because they think you want to buy a food product.

Strictly speaking, with every communication there is always an expansion of the meaning space. Such an expansion can only be achieved by adding a second communication channel, as shown in Figure 1.2. In human communication, this second channel is always present due to the situational channel that is always present (see Figure 1.1). We therefore live in a world in which we constantly expand our meaning space through our five senses and the situational channel.

We discovered that without knowing the name of the plant you would never have been able to purchase it. On the other hand, you also have to be in the right situation, namely in the right store, so that the communication is successful and the appropriate information arrives.

If we now apply this knowledge to our communication model, we come to a first expansion of the model.

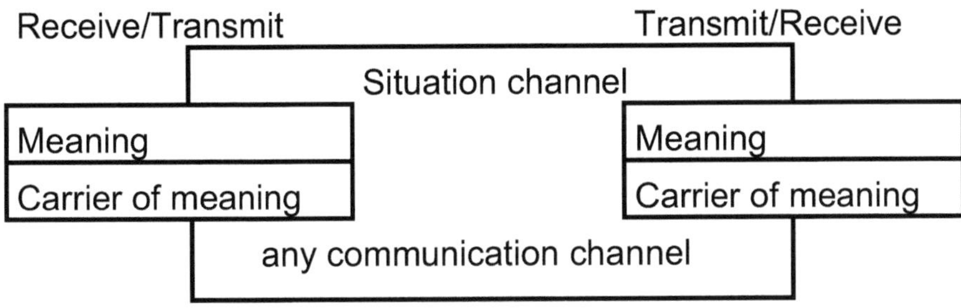

Fig. 1.4: extended communication model

In every communication process, in addition to the active communication channel, there is always a second channel that also determines the information transmitted. This has been referred to as the situation channel. It represents the circumstances under which the sender and receiver communicate. In this way, it also represents the already existing spaces of meaning of the sender and receiver.
Only when the situation is appropriate and the correct meaning is exchanged can one speak of successful communication. Successful means that the meaning of the sender matches that of the receiver.

We have arrived at a dynamic model that no longer needs the prerequisite of identical spaces of meaning and can therefore now reflect both the creation and use of identical spaces of meaning. We have to realize that the information is no longer determined by just one channel, but is always influenced by the situation channel. With the one exception that the meaning spaces are already completely identical. (E.g. in digital communication between computers) A changed situation usually also changes the transmitted information.

The purpose of developing models is to represent a process in its most elementary form in order to promote understanding of more complex processes and to be able to define terms more precisely.

In fact, it is a typical scientific approach to define terms *precisely* based on available models in order to achieve a better understanding of the terms. Let's start by defining the term meaning.

Meaning

Meaning is what is to be transmitted by the sender or received by the receiver and is therefore assigned to the corresponding meaning carrier. The meaning carrier is assumed to be identical for the sender and receiver.

(There are cases in which the meaning and meaning carrier are identical. In this case there is no need to assign them! We will come back to this case later.)

Information

Information is exactly the meaning that the recipient assigns to the meaning carrier.

This is a very general definition. It makes no statement as to whether the meaning received corresponds to that intended by the sender. In practice, both cases occur,

a) in which the correct meaning is assigned to the carrier, and case
b) in which an "incorrect" meaning is assigned.

We have thus moved away from the technical definition, which only allows case a). But we should highlight the importance of case b) here already, see EPR-Experiment in the next chapter.

An example: The moment you go to the store near the jungle and ask for Haldala, an exchange of information actually takes place. Despite the differences between your meaning space and that of the seller, there is at least one thing in common. This is the connection of the name "Haldala" (meaning) with the plant you eat (meaning). The purchase works (case a.), you get the plant you want!

On the other hand, you can easily imagine that if you go to a store in a completely different country and ask for "Haldala", you might naturally get something completely different than a plant. In this case (b) the meaning spaces were not identical; the seller assigned the term "Haldala" a different, but still a meaning. Communication was not successful!

Communication

Communication is the creation of information and the attempt to successfully transmit meaning.

Using those three definitions, shows that the meaning becomes information precisely during the act of communication. Meaning is what appears to us as information and information is what appears to us as meaning.

As we can see, the two concepts of information and meaning are very similar. But it is still useful to be aware of the subtle difference. This allows us to distinguish between a static and a dynamic aspect. Information is what only occurs during communication, while meaning also exists before communication with the sender or remains with the receiver after communication.

After communication has taken place, the recipient's space of meaning has been expanded by information. The information then disappears and becomes a meaning in the recipient's space of meaning.

Information is therefore the difference between the meaning - space of the sender and the meaning space of the receiver in relation to a very specific situation $I=MSS(S)-MSR(S)$.

From the sender's perspective, communication was successful if the meaning sent matched the meaning received.

Once again: There is no information in itself, but rather information always arises during the act of communication. From this it can be deduced that information is in principle independent of the meaning carrier. This brings us to the following important findings:

Information is neither tied to a (material) carrier nor influenced by it, but is determined exclusively by the act of communication. Information is, in the truest sense of the word, a metaphysical entity.

Meaning is always tied to a carrier or, in the borderline case, identical with it. In this case, it is no longer necessary to distinguish between information and meaning.

If we take our consideration to the borderline case that meaning and carrier are identical, we arrive at a worldview in which everything consists only of meanings or information. As we will see in Chapter 2, this view is very close to the view of modern physics.

The further scientific approach will be, applying the communication model developed, in order to arrive at new ideas about reality. How exactly these ideas correspond to reality - remains a question that still needs to be answered. If necessary, the present model must be further modified or expanded. However, with the aim of retaining the basic concepts.

The hope that there is a model that is arbitrarily accurate and accurately describes reality has occupied many researchers for a long time and many still believe in this illusion today. For a long time, this illusion was nurtured by the introduction of so-called "closed systems". Closed systems are systems that do not interact with the environment in any way. The aim was to minimize the influencing factors and achieve an exact description of these systems.

Upon closer inspection, it can be shown that the introduction of information into the scientific world ends the existence of closed systems.

The moment I receive information from or about a system, the system is no longer a closed system; an interaction has happened with us, however indirect this may be.

On the contrary, in atomic physics it was found that the isolation of a system was disruptive when describing it. Many calculations required integration over the entire (i.e. infinite) space, which in turn requires the openness (non-closedness) of the system under consideration.

We will come back to this fact in connection with the consideration of entropy, because it indicates that *there is no model that is arbitrarily accurate!*

We now want to make further definitions that will later lead us to expanding the communication model. The well-known term "consciousness" should be placed in a new context.

Consciousness

Consciousness refers to the entire communication process including sender and receiver. Consciousness is a self-referencing mechanism. The expansion of meaning spaces that occur during communication is identified as the expansion of consciousness.

It is clear to us that these definitions do not directly correspond to those traditionally used, but it has the enormous advantage of being scientifically accurate. This means that the concept of consciousness has a clear context of meaning and can therefore be communicated precisely!

With the help of the above definition of terms, we will expand our communication model into a knowledge model respectively a consciousness model in the following 2nd chapter. In order to further improve our understanding of scientific methodology, the history of science is briefly presented below using physics as an example. But this also leads us to the important findings of modern physics, which we will then integrate into our model of consciousness.

2 History of perception

The history of perception is very old. For me it leads to the history of physics and later to the history of love.

The history of physics in my presentation is sometimes incomprehensible in detail for non-physicists. I would like to point out that general understanding is often more important than following each individual step.

The Tree of Knowledge

Perception is a special form of communication (Fig. 1.4). The perceiver is the receiver and the outside world, i.e. what is perceived, is the sender. The perceiver has a space of meaning which we want to call MS S (S=subject). What is perceived also represents a space of meaning MS O (O=object). According to our communication model, the objects that confront us must already have a meaning (or many?), because otherwise communication, i.e. exchange of information, with our environment is not possible (see Figure 2.1), we can't recognize them! Communication takes place via the five different channels of our sensory organs (eyes, ears, mouth, nose, skin) and the situation channel.

Let's imagine ourselves for a moment in the perception of an animal. We notice that the animal's perception is still immediate and unbroken. This means that the animal's space of meaning is largely identical to the space of meaning of its environment. The communication between the animal and its environment is not yet overlaid by a meta-perception (copy, see Figure 2.1) and therefore there is neither a difference between meaning and carrier nor between animal and environment. There has not yet been a separation between one and the other part of the whole.

The Old Testament teaches us the transition from animal to man, the fall of man: When man ate from the tree of knowledge, he was expelled from paradise.

In addition to the original space of meaning, humans built up their own new space of meaning based on their ability to reflect. We formed a copy of the perception (see Figure 2.1). The prerequisite for the ability to 'know' came about through the inner information channel.

The person later recognizes themselves and their environment in the copy and draws their self-image from this copy. The ego identifies itself with the copy. In this cognitive process, the ego experiences itself as a being separate from the rest of the world (expulsion from paradise).

As shown in Figure 2.1, this knowledge represents an internal communication process that we will call self-reflection.

Interestingly, the emergence of self-reflection goes hand in hand with the loss of the person's identity with the whole.

In the initial stage of human history (stage between animals and humans) he does not yet recognize this process, he is not aware of the separateness, there is no elaborate self-reflection yet. Therefore, the Bible teaches him exactly this process (fall into sin), which is supposed to give people self-confidence. Later he learns to perceive his separation from the rest of the world; he recognizes himself as a perceiver. The 'I' came into being.

The copy of the perception process influences the perception process itself. The copy appears to the person as incomplete. This gives him the impression that his surroundings have lost their importance. The meaning of what he perceives is no longer immediately clear to him, as he now sees it through the lens of the copy. He believes that he does not know what he perceives, although only his image (copy) of perception is incomplete. The copy does not contain all the meanings that already attach to the objects in the outside world. Man was neither aware that what he perceives depends on the perceiver, nor that the perception process itself is influenced by its own copy and vice versa.

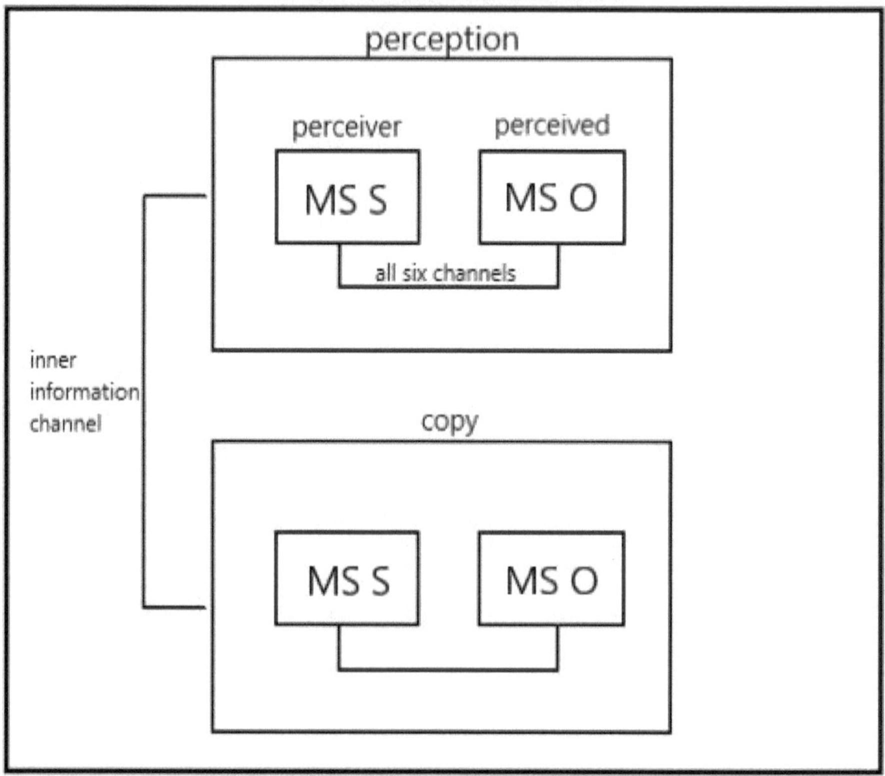

Fig. 2.1: People perceive themselves as perceiver,
cognitive faculty and self-consciousness emerge.

The perceived

Interestingly, in our culture 'what' is perceived becomes the subject of very close investigation, rather than the perceiver or the perception process itself. There have been Eastern cultures in which the process itself was the primary focus. These then gave rise to Taoism, for example.

Our natural sciences have carried out sufficient investigations into what we perceive over the past centuries. In the following discussion of the development of physical research, we will find that this path of knowledge leads to the inclusion of the perceiver or the perception process and ultimately leads back to unbroken perception.

Let us now consider the evolutionary path of knowledge in detail. People began to examine what they perceived first. What is perceived can obviously be further divided into individual objects (= new, smallest common space of meaning). These objects can be named (identity of the meaning spaces is established). This was the first great task of the biblical Adam. He had to find a name for everything that surrounded him.

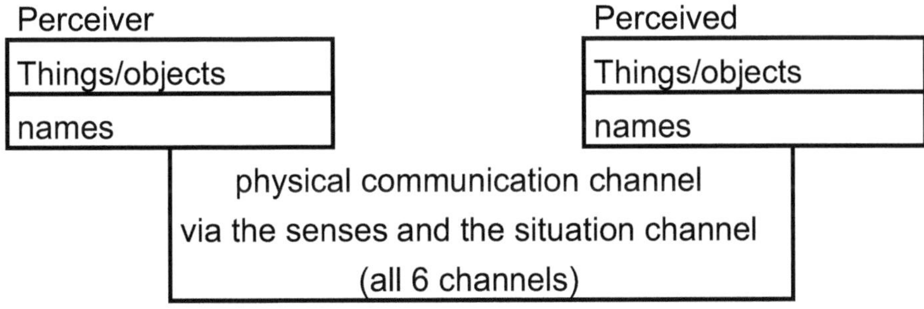

Figure 2.2: Task of the biblical Adam

The distinction between humans and the environment (fall of man) ran parallel to the distinction between meaning (things/objects) and signifiers/carriers (names).

This brings us to the beginning of the history of natural science and physics. The first major achievement to be mentioned here is the introduction of counting objects and thus the formation of groups of objects. There is one sheep, but there are also two sheep, or three, or the multitude of all sheep. This was a huge step towards abstraction. A first tool had been found to make oneself independent of the names of the objects. One could now speak of the number three. Whether these were three sheep, three horses or three people was entirely up to interpretation. In the next step, we became independent of the specific number and introduced the so-called variable that represented this number. From this, systems of equations and functions developed and thus essentially the mathematics we know today. In the meantime, no one was idle when it came to examining the objects themselves. They looked at the properties of the objects and found that these properties could also be divided into groups. Organic substances, metallic substances, volatile substances, earthy substances and so on were found. In this way, people moved further and further away from the original reality. Because now reality was more and more what people had thought up in terms of categories and formulas. Consequently, the categorizations became more and more precise and accurate, and it seemed as if nature could soon be fully described. A big step was the discovery of the atom model. All substances surrounding us could be classified and, it was hoped, broken down into their smallest parts. It was believed that we had found the smallest parts of the objects around us and that we could use chemistry to explain the causal

Love and Physics

relationships between these smallest parts in order to find a *complete picture of reality.* As long as you followed the simple model shown in Fig. 2.2, nature was exactly as you imagined it to be.

The linear, causal thinking corresponding to this idea brought with it the concepts of past and future and thus the course of time. In this way of thinking, time was reversible, or more precisely, the processes taking place were reversible and traceable back in time. The study of complex systems containing many, many atoms led to the concept of entropy. Entropy is a measure of the disorder of a system, or the irreversibility of the processes taking place in the system. An example:

1. Imagine that a thermos flask is half filled with hot water and then closed. We neglect the heat that now escapes from the jug and therefore assume that the jug is a closed system. As the inside of the jug moves towards thermal equilibrium, maximum disorder will now arise in the jug. This means that water vapor is distributed evenly in the second half of the pot. Since the pot is supposed to be a closed system, this process is not reversible; the entropy has increased irreversibly.

2. Now let's imagine that the jug above was cylindrical and in the lid of the jug there was a piston with a spring that releases itself as soon as thermal equilibrium is reached in the jug, that means the entropy is constant again.

The piston would now compress the water vapor, causing the entire system to heat up. This heat energy would then expand the water vapor again and push the piston back all the way into the lid. A reversible process has taken place, the system has returned to its original order, the entropy has remained constant. One must always remain aware that there should be an ideally closed system that does not exist in reality. Also, the friction in the spring should be neglected. Only then will the second example work! So actually, it wouldn't work or only work close to it.

The entropy of a closed system is either constant, in which case there is a reversible process, or it increases, in which case the process is not reversible.

It was recognized that most processes in closed systems are not reversible. These processes are complex, non-linear processes. (our first example) From this fact it was deduced that natural, closed systems reach a maximum of entropy, i.e. maximum disorder, after any length of time. Under the assumption that our universe is a closed system, the heat death theory was derived. It says that after a certain time our universe must reach a maximum of entropy or a maximum of disorder. From this in turn it was concluded that time must have an excellent direction, namely towards this disorder, i.e. into the future.

The idea that our universe is a closed system was questioned early on by chaos theory. Chaos theory deals with chaos or the spontaneous emergence of order, i.e. the formation of similar, defined states. The chaos theorists determined that a prerequisite for the emergence of order is the 'not closedness' of the systems under consideration.

The best example of the emergence of order is the emergence of life. Living structures have a high degree of order and arise in open (not closed) systems. Through this creation of order, life itself counteracts the increase in entropy, it reduces entropy. Only *not closed systems* in which life arises ensures compliance with the law of entropy. The law of entropy (second law of thermodynamics) prescribes the following: Processes in which the entropy decreases only take place in *not-closed* systems.

Since the law of entropy and the fact that there is life in our universe are supposed to be general, we conclude *that our universe must be an open, not closed, system!*

The findings described represent only half the overcoming of the mechanistic (linear) world view. The development of physics at the beginning of the 19th century ushered in the complete overcoming of the linear world view. In 1905, Einstein published two papers that were to thoroughly shake the existing edifice of physics. Both the special theory of relativity and the postulate that light is particle-like (quantum hypothesis) triggered a revolution in physics at the time.

The special theory of relativity assumes that the speed of light is constant in all conceivable reference systems and represents the greatest possible information transmission speed. This gave Einstein completely new possibilities for making statements about space and time.

Space and time were no longer independent as before, but were now firmly linked to one another. Two observers who are at different locations in space each have their own valid time. Or, at any given time there are many different realities, depending on where in space you are. The dependence of space and time was the first major shock to the world view of the time. Until now it was assumed that space existed completely independently of time and vice versa. But things got much worse.

Once again it was Einstein who provided the impetus for the later collapse of the entire physical world view that had prevailed up to that point. His second work states that particle properties (quanta) must also be assigned to an electromagnetic wave (light). It was precisely this "duality" of light that was the basis for an entire team of physicists to develop quantum theory in the 1920s and 1930s. Physicists like Heisenberg, Bohr, Schrödinger, Dirak, to name just a few, had to deal with the fact that the electromagnetic wave had wave properties on the one hand and particle properties on the other, depending on how you looked at it. If light was viewed as a wave, the typical wave properties were observed, such as diffraction, interference and so on. However, if you looked at light as a particle/quantum, quantum effects also became visible. Here, for the first time, the effect was that *the objects to the observer were no longer - independent of the observer (perceiver).* Depending on which model the observer had of his object, the object behaved differently. Events now accelerated.

The recently discovered elementary particles neutron, proton and electron, which were for some time considered the final basic building blocks of matter, also fell victim to this duality. The electron was now no longer just a particle, but could also just as easily be understood as a wave. The same was true for neutron and proton. Due to the wave properties, the particles lost their exact location. They were distributed over a specific spatial element. It was Werner Heisenberg's great achievement to mathematically quantify these uncertainties in the subatomic range. His equations, known as uncertainty relations, mark exactly the limits of the describability of nature. This was, so to speak, the last big blow that caused the physics of the time to fall into a complete vacuum. The world could no longer be described exactly.

Matter no longer exists with certainty, but only shows the tendency to exist. Events only have a tendency to occur. The traditional idea that one has a reality that can be described "objectively" had been completely destroyed. The physicists therefore had to come up with new mental models and ideas. The physicist and successful author Fritjof Capra describes a first approach as follows:
"So subatomic particles are not things, but connections between things, and these things are in turn connections between other things and so on. In quantum theory you never get to things. You are always dealing with webs of interrelationships."
This approach was later expanded to state that the web of interrelationships encompassed the entire cosmos. That means everything interacts with everything else. *Every event is - influenced by the entire universe and vice versa.* (see *entanglement* for further reference)

As part of the development of the new ideas, even Einstein's postulate, according to which the speed of light is the greatest speed of information transmission, had to be put into perspective.

Einstein himself devised a thought experiment that would support his postulate. From this experiment, which later became known as the Einstein-Podolsky-Rosen (EPR) experiment, John Bell derived a theorem according to which an interaction can take place instantaneously in a certain physical system, that is, there is no time for the effect to occur required, even if the interacting partners are many thousands of kilometers apart. (=entanglement) Under certain conditions, such a system can also transmit information (in the non-technical sense, see Chapter 1) at infinite speed, apart from the reaction times of the transmitting and receiving devices. With the implementation and confirmation of the EPR experiment, Einstein's postulate was refuted.

It is time to take a closer look at the difference between the concept of information defined in this book and the technically standard concept of information. There are obviously ways in nature in which objects can interact without transmitting any information in the technical sense. The interaction of objects is even of a causal nature. If the state of the physical system changes in a certain way here, then it also changes in a very specific way in the system many thousands of kilometers away. According to physical experiments, it is not possible to transmit information of a technical nature via this interaction. We remember that the technical concept of information means that a predetermined meaning must be transmitted exactly. This in turn means that parts of the meaning spaces of the sender and receiver must be identical. This means that the spaces of meaning must be coordinated with one another. Nature obviously allows an exchange of information that does not require previously agreed upon spaces of meaning. This means that the information does not correspond to expectations (case b) on page 20). We will discuss this type of information transfer in more detail in the next chapter.

Interestingly, Einstein had provided the physical basis for the above result: Einstein discovered the dependence of space and time. Figure 2.3 shows Einstein's space-time continuum. For reasons of representation, the space is represented here by a single axis. We get a two-dimensional representation of the space-time dependence. The coordinate origin represents the now, while the timeline upwards shows the future and downwards the past.

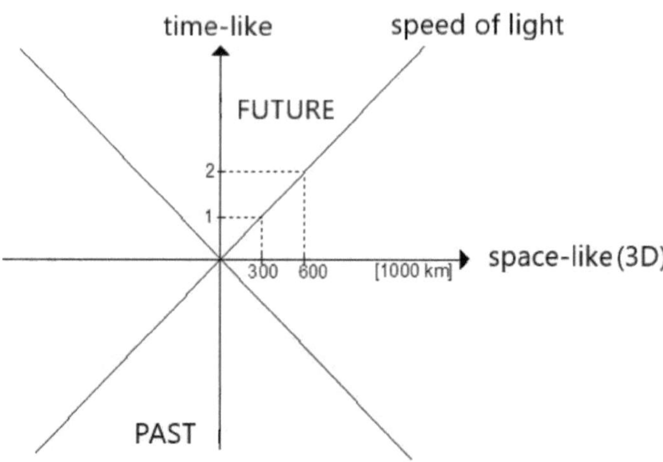

Fig.2.3: Einstein's space-time continuum

According to Einstein, there is now a highest speed in this space-time at which objects can interact. This is the speed of light at approximately 300,000 km/s. The line drawn represents exactly the positions in the space-time continuum on which all light in our world moves. Every ray of light that we emit in the now runs exactly on this line, but the prerequisite is that the light is in "free space", i.e. not in any medium (e.g. water or air).

Einstein's statement was that the world we can experience lies exactly between the lines of light, i.e. in the time-like area. In this area we find all effects that propagate at a speed that is *less than* or equal to the speed of light.

From this Einsteinian model of the world, it is possible to derive, at least theoretically, an effect that propagates at a speed *greater* than that of light. All events occurring in the space-like realm of representation could interact with our "now" at a speed greater than the speed of light. And this is exactly what was required in the EPR experiment (see above): If our measuring device is at the now point, the partner to be measured is somewhere on the spatial axis. All events that can expand at infinite speed lie on the spatial axis.

What Einstein did not know, and what we do not know either, is in what medium these effects propagate in order to reach infinite speed. We know that it is certainly not our "normal" space. This is an interesting research area for modern science.

It was discovered that not only does everything interact with everything else, but that this interaction takes place at infinite speed. *Nature studied by physics turned out to be an indivisible whole.*

Physicists got used to the fact that there were no longer any uniform views, but that one had to accept the validity of different views. It was found that the structure of the object of study appeared to be very similar to the structure of the mind of the examining observer. Capra expressed it this way:

"The fact that all the properties of particles are determined by principles that depend closely on the methods of observation would mean that the fundamental structures of the material world are ultimately determined by the way we see this world; the observed structures of matter would therefore be reflections of the structures of our consciousness."

The principle of causality

What constitutes our thinking cannot actually be described. If you think, you think, then you only think you think, because you think that you think that you think...

This is certainly just a small example of the limitlessness of our thinking. Our thinking is a typically open system. We will therefore initially limit ourselves to the part of thinking that serves us for communication. We will limit ourselves to the part that is common to all people when thinking. We need a lowest common denominator of thinking, so to speak, to which every person would immediately say: Yes, I think that too. A body of thought, a theory, a philosophy that is based on this common denominator has the greatest chance of being understood and applied by all people.

The smallest logical unit that people can agree on today is the principle of causality. Sentential logic and thus all of mathematics are based on this if-then principle. What appears to be a convention of thinking in mathematics is actually based on the experience of elementary natural events. I would therefore like to explain the causality principle as a cause-effect principle using a real-life event.

Dear reader, imagine that you are once again in the jungle and standing under a tree. Suddenly a fruit falls on your head.

Of course, when considering this process we have to make some simplifications. We assume that you have been under the fruit of this tree for quite a while. We can then say that when the fruit falls, it hits you. Or, more abstractly, if event A occurs (fruit comes off the tree), then event B follows (fruit hits you). We have exactly one cause A opposite one effect B. Of course, we can now create chains of such events of any length. A light wind comes up and shakes the branch of the tree, causing the fruit to fall loose. She falls on us, we fall over. Our companion carries us to a hospital...

Let's do a thought experiment in which we imagine a closed system in which there are only three events. Event A (the fruit falling) is the only cause. The second event B (you are hit) is the first effect and the event C (your companion is hit) is the second effect. You are standing right under the fruit, so A (the fruit falls) becomes the cause of event B (you get hit). Your companion will never be hit because he is not standing under the fruit. So "from A follows B" is a true statement and "from A follows C" is a false statement. We have restricted our system so that the statement "from A follows C" *must always be false*. We have come to an exact description of this artificial system, which makes the system's behavior 100% predictable. For we can predict that if A occurs, then B will occur and C will not occur. We see here that causality also imposes the concept of time on us. Because event B is something future in relation to event A.

Of course, in our real life the situation is completely different. We can easily imagine that just before the fruit falls, your companion takes your place, or a gust of wind moves the fruit towards your companion, or, or, or... There are endless possibilities, i.e. cause-effect chains, who can have an influence here. The statement whether you will be hit is in fact indeterminate until the event occurs, i.e. true or false.

The difficulty with an open system, as we have seen, is that we may not know in detail the causes that come into effect, and therefore this system is unpredictable. Subjectively speaking, the reader will remark, I have no difficulty with this. Because I don't care about the reason why I choose to eat a banana instead of an apple when I'm hungry. It's only when I try to tell someone why I'm eating the banana and not the apple that I might get into trouble. Or to put it more extreme, if someone else tries to predict whether I will eat an apple or a banana, they have the greatest difficulty. But that's exactly what we want. We want everyone to be able to predict the behavior of the system under consideration. In order for everyone to be able to make these predictions, our model must be communicable. We must be able to communicate it to the person who wants to make the prediction.

We have already seen the worst-case scenario in the history of physics, in which our causal models collapse and the communicability of the models is jeopardized. It is the classic example, the physical phenomenon of light (electromagnetic waves vs. particles). Physicists have come to the point of using two different, contradictory models to explain one and the same reality. Scientists have always strived to find a uniform, consistent model. However, the developments of modern science show more and more clearly that this is the wrong goal. Rather, it must be recognized that many different, contradictory models describe reality in fragments.

The dilemma becomes even clearer when we apply our considerations to the models themselves. It turns out that even models cannot be concluded without getting entangled in contradictions. The mathematician Gödel discovered a law that states that complex mathematical systems of axioms allow conclusions that lead to theorems that can no longer be proven mathematically within the system. Nevertheless, these sentences can be recognized by humans as true, i.e. valid sentences. This is because we are outside the axiomatic system.

You can also formulate this differently: In a logical, syntactic system, freedom from contradiction is fundamentally not achievable as long as you stay within the system (requirement for closure). This peculiarity of mathematical and linguistic logic confirms our considerations above. Even if we move exclusively within a logical, mathematical model and complete it, we cannot achieve freedom from contradictions.

So, we find that nature must always be viewed as an open system. In an open system there is no final cause to which all others can be traced back. Because this cause has another cause, etc. This means that the open system loses its complete describability.

Science is now in serious difficulties because, on the one hand, it demands that nature be describable so that the models of it can be easily communicated and understood. Therefore, it assumes causality and closure in nature and forms causal, closed models. On the other hand, it requires that these models be complete, accurate and highly predictive.

This is a fundamental problem in the world of thought, perhaps the most fundamental problem of all. The statement "God is all that is" reflects this problem very well. This statement is to a high degree **C**ausal, **C**losed and **C**omplete. We have a perfect model (triple C), but the statement contains almost no details and therefore has no predictive power. (In the sense of our causal thinking)

The only way to overcome this problem is to view all systems as fundamentally open and to give up the complete, uniform describability of nature.

Eastern mystics speak of the eternal karmic wheel as an infinite chain of cause-effect relationships that we are constantly producing and changing. Two reasons are repeatedly given for the karmic entanglements. On the one hand, the deceptive image of reality that is created by our language and on the other hand, our freedom to choose. When making decisions, we necessarily base our decisions on an incomplete worldview constructed through language.

This results in two approaches to stopping the karmic wheel: 1. no longer making decisions, i.e. no longer choosing and 2. no longer communicating via language. Both approaches interestingly have been demonstrated again and again by Eastern mystics: the communication aspect is eliminated by withdrawing into a hermitage or taking a vow of silence. Choosing is avoided through unconditional surrender to one's fate.

We can derive our "free will" from what has been said so far. Free will is the instrument that creates new causes in existing causal systems and thus possibly also creates new effects. "Free" just means that we are not in a closed system. And it does not mean that we leave our causal thinking, because that is impossible if we want to communicate.

So, there is an open system in which external intervention occurs. Since we ourselves also move in a causal world (suppositional), the entire system that is put together should also be causal, that is, deterministic. Nevertheless, we intuitively hold on to the claim that our will is free. And we are right to do this because, as we have discovered, there are, in principle, no closed systems in nature.

This means that there are *an arbitrary number of* influences that affect us, and our decision is therefore truly free. The will is free in a causal world!

Of course, the world is only causal because we still need causality to communicate today. Our collective thinking today is necessarily causal and therefore our common understanding of the world is causal.

In this context it becomes clear that there are also other possibilities for overcoming causality. By developing a type of communication that is not aimed at theoretical completeness, but rather takes people's practical life situations and the mystical aspects of communication into greater account. The corresponding communication model is developed in the following 3rd chapter.

As an example, I would like to mention the "Findhorn" community. Their original community demonstrated how they could collectively cultivate fruit and vegetables that have defied all conventional scientific explanations to this day. Despite unfavorable conditions, the plants grew to a supernatural size and splendor. This is an achievement that goes beyond any physical theory, but as can be expected, it can only be communicated/understood to a limited extent. But better local miracles than global suffering.

The quality of knowledge

The results of modern physics shouldn't surprise us, because we would have arrived at these results if we had applied our perception model (Figure 2.1) to the external world from the start. In order for cognitive perception to take place, the space of meaning of the perceiver (MS S) must be identical or at least similar to the space of meaning of what is perceived (MS O).

For example, we imagine performing an act of communication with the subatomic world. So, we choose some medium, for example a particle accelerator, and try to establish contact with the subatomic world. In order for information to be obtained, that is, a meaningful result can be achieved, we must assume that the subatomic world has the same or at least a similar 'space of meaning' as we as the observer have.

From the mere intention of learning about the subatomic world it follows that it must be as we imagine it to be. Of course, this idea is constantly expanding, and with it the properties of the subatomic world are expanding.

Consciousness has the property of constantly expanding, which is why physical theory can never be complete. It will continue to evolve over and over again.

Many physicists will therefore set out to achieve a new quality of knowledge, to return to the unbroken view of nature that lies beyond all models and theories. The aim is to return to direct, intuitive knowledge, which can only be achieved through the *personal development of each individual.*

I would like to point out that the communicated findings of individual scientists, or those of an entire community of researchers and scientists, can provide us with assistance, but only everyone can walk the path of knowledge themselves. Depending on what significance the scientific teaching has for the individual, he or she feels more or less affinity and is thereby more or less strengthened on his or her path.

I'm therefore trying to build some bridges on this path. As is almost always the case when building bridges, the material the bridges are made of comes from the shore from which we come.

Love

Physics attempts to complete the image or copy of perception (see Figure 2.1) by examining what is perceived and with the help of the internal information channel. At the same time, it attempts to eliminate the difference between the perceiver and the perceived. Nature turns out to be more and more an indivisible whole. The self-consciousness that has arisen thereby strives towards its perfection.

The immediate continuation of this development is determined by a new quality of knowledge. This is the perception of the wholeness of the world through love. Here the new, self-consciousness is contrasted with world consciousness and at the same time the unity and equality of both is recognized through love. Divine conscious-ness arises.

If knowledge of what is perceived (physics) was the integration of the perceiver's inner world, love is the integration of the outer world as a whole.

Both processes run in parallel (Figure 2.4), so that love is not possible without knowledge and knowledge is not possible without love! Finding the identity of the perceiver and the perceived is the same as finding the identity of ourselves with the rest of the world. We will return to the relationship between knowledge and love in the next chapter.

Interestingly Jesus Christ taught us love. He taught how to overcome the separation of man from the world. The separation between man, the world and God is overcome by bringing God back into the world. The lived love for God (the Universe) overcomes the boundary between man and the world, it allows man to become one again with the world and all that is and frees us from original sin.

Our self-consciousness communicates through love with the world consciousness, i.e. with everything outside of ourselves. This creates divine consciousness. Of course, divine consciousness also arises when we love another person. People recognize themselves in others. He recognizes the identity of himself with the other.

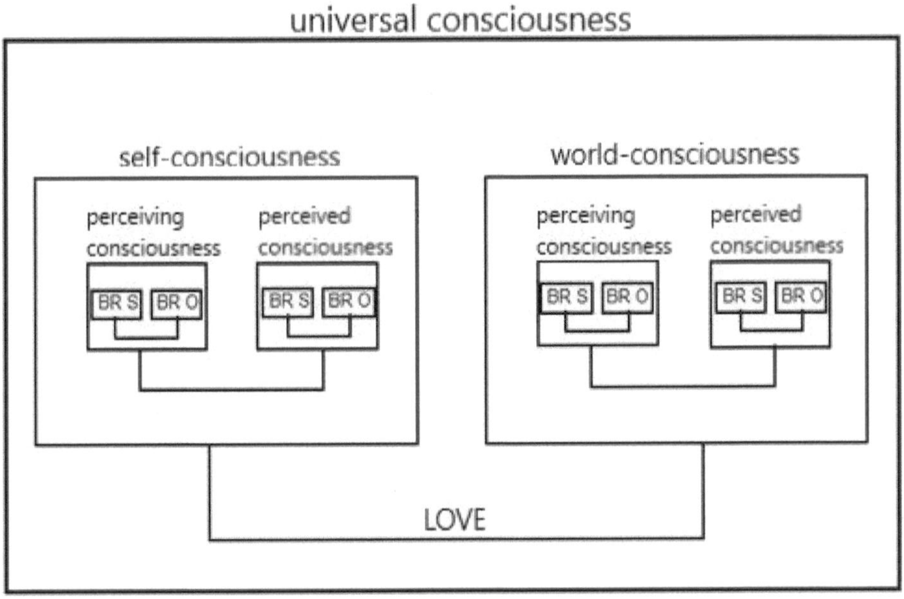

Figure 2.4:   Man recognizes himself as a unity with the whole.
Love and divine consciousness arise.

Love goes far beyond our scientific world view. It forms, so to speak, the link between the scientific and magical worlds. As we will see below, there is no separation between these worlds, but rather they are causally related to each other.

3   Consciousness

Spatial communication

We would like to first summarize what we find important in the first two chapters:

There is space- and time-like interaction. Everything is connected to everything via space -like interaction, which means that everything is contained in everything, or to put it another way, there is only one indivisible whole. From this knowledge it follows that any two objects in this reality have the possibility of interacting with one another. They are always connected to each other through some medium. It follows that, wherever the objects are in the space-time continuum (including objects that lie in our "future"), they in principle have the possibility of communication, that is, of exchanging information.

We will take the crucial step for this book of expanding our communication model to include the above aspect and arrive at a model with which the already defined concept of consciousness can be described even more precisely. We will introduce a space-like communication channel that complements the existing time-like channel.

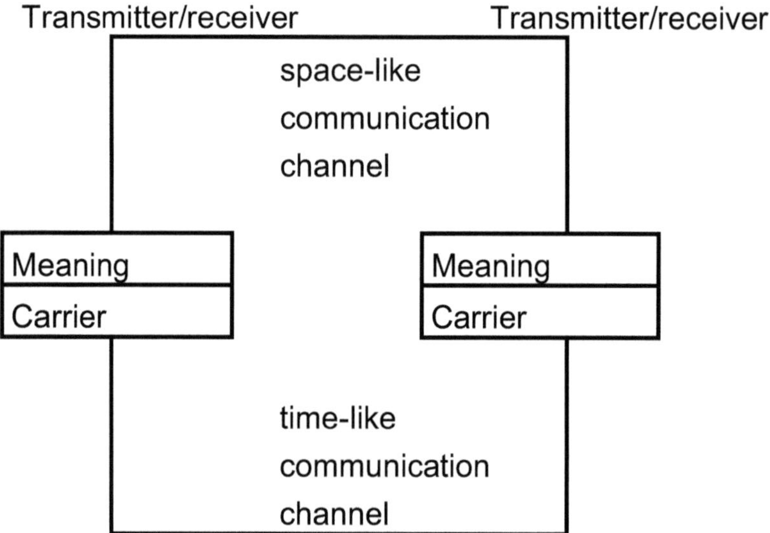

Figure 3.1: Consciousness

In doing so, we have taken the results of modern physics seriously and introduced a channel that connects everything to everything in an arbitrarily short time. This channel is effective in all communication, including our perception.

We now come, for the first time, to an assertion that cannot be directly verified historically, physically or logically:

As humans, we have the ability to constantly access this space-like channel through our inner perception.

Internal perception here means the perception that does not occur via our external sensory impressions, but rather that which is possible for us via ideas, thoughts, inner voices and images.

If we turn inward, most of us will notice our own thoughts. There are different levels of inner perception. Initially, the mind and rational thinking dominate perception.

But the more we relax (e.g. when falling asleep), the more images and fantasies, including daydreams, will come to the fore in our inner perception.

Our ideas, fantasies and dreams are largely determined by the space-like channel. Through these inner space-like channels we are able to create connections that we are not yet aware of at all.

How does this inner access work? A good access method is to close your eyes and try to shut off all perception. The work of the thoughts should also be completely reduced so that the inner perception can concentrate entirely on nothingness.

What happens when we communicate with nothing? Let's apply our model of consciousness (Figure 3.1). After switching off all 6 time -like channels, the information received is only determined by the space-like channel. The information received will automatically appear in the form of images or thoughts that we know, since it can only find expression through our own space of meaning (everything that makes us human). There is a complete reflection of ourselves. We find nothing other than ourselves.

Our self contains the world consciousness, but according to our model nothing more than our self can be known. At this stage, world consciousness has a part that is unrecognizable to us.

Only when our own space of meaning has become nothing do we have a direct and immediate experience of unity or God. If nothing communicates with nothing, our model no longer allows any statement. We can then only experience or live the unity with everything that is. We thereby go beyond the human form. We are then no longer human in the traditional sense; we have become something different.

If we look at the expanded model of our perception (Fig. 3.2), we see how love and divine consciousness find perfection. Through the awareness of omnipresence (space-like channel) via inner perception, man now feels unity with himself. In the final consequence he experiences that he is identical with everything that is, that every consciousness is identical with the divine conscious-ness. In completion, all the various stages of the model collapse and the original communication ceases to exist. We then come to a really simple model (Figure 3.3), which of course only contains a few details, but because of how it was created it says it all.

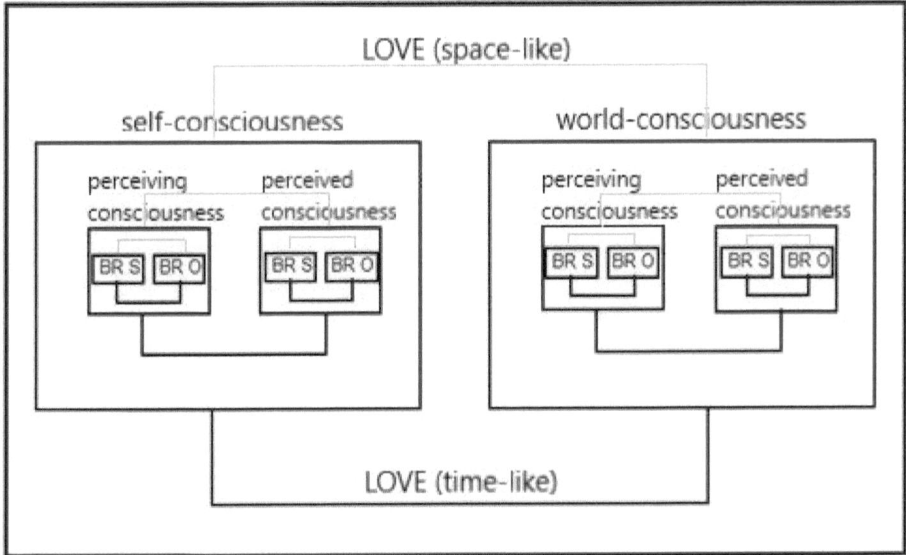

Figure 3.2:   People feel unity with the whole.

Love and divine consciousness are complete.

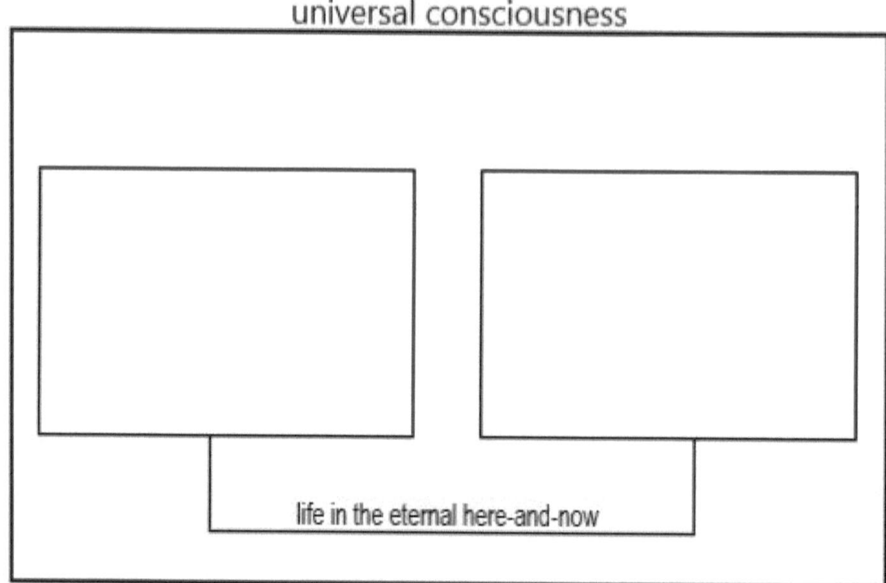

Figure 3.3: People live unity with the whole, they live in the eternal here-and-now

There are already many people and organizations who have started verifying the above claims. We will further deepen the practice of inner space-like communication in Chapter 4. But first we work out a few conclusions from our previous findings.

Individual reality

The impact of our model of consciousness on our general - understanding of reality can hardly be overestimated.

Reality, as physics has recognized and our model of consciousness shows, is made up of something that takes on exactly the properties that everyone assigns to it. There is no reality in itself, but everyone perceives something that they

Love and Physics

believe to be real. There is not one *objective* reality, but there are *many* realities. This does not mean that the realities are not similar. On the contrary, similarity ensures the continuity of our perception.

We become aware that we have a consciousness. This allows us to participate in the constant creation of reality. When we change our consciousness, reality also changes.
Reality is not the sum of all human actions and ideas. There is no ONE common reality. There are just as many realities as there are consciousnesses. Every consciousness has its own reality. This also follows directly from our model: What we perceive is directly determined by our own scope of meaning and this varies from person to person. *We have an individual consciousness.*

The linking of the spaces of meaning through the spatial communication channel leads to similarities in realities. We can therefore speak with some justification of a shared reality.
But this shared reality is something far different than what is generally understood by "reality". As long as we ignore the space-like channel and do not include it in our thinking, "reality" will seem contradictory, complex or even chaotic to us.

Although everyone has an individual reality, it still affects the reality of all other individuals due to the space-like connection of everything to everything else. An infinite number of realities exist side by side at the same time.

Every moment of the present therefore fundamentally changes the world. Every thought and every action of every single person determines immediate world events. The very personal influence on world events cannot be overestimated. If my actions and thoughts change, I change. If I change, everything in the world changes. We are personally in a position to determine what happens in the world. Every person does this all the time because *every person has an individual consciousness.*

Emergence of All That Is

Since EVERYTHING is contained in everything, we can take any object as meaning space 1 (MS 1) in which EVERYTHING is contained. Since a space-like connection means an interaction with infinite speed, something like an OMNIPRESENCE can be thought of, which enables the introduction of a second completely identical meaning space (MS 2), which is connected by a space-like communication channel.

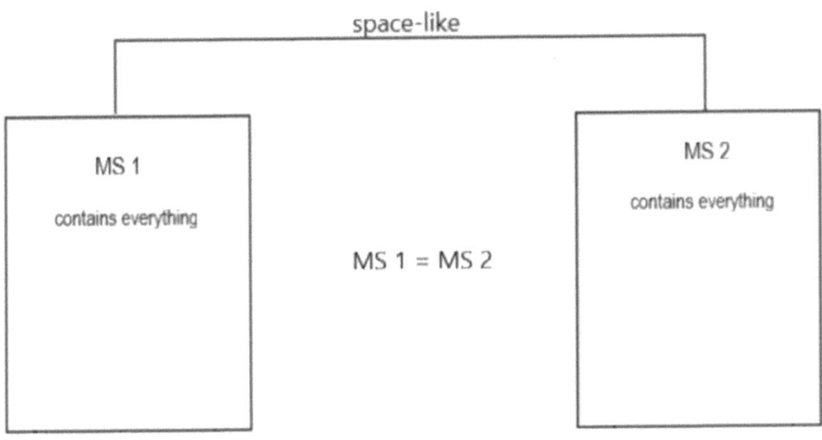

Fig.3.4a: The eternal here-and-now, 1st phase

The division into MS1 and MS2 is initially purely theoretical; we still have one unit before us, the all-unity. MS1 contains MS2 and vice versa. This construction of two spatially connected permanent identities (MS 1 = MS 2) initially seems pointless, since communication has become an end in itself. There is no information that the other identity does not already have. In fact, universal consciousness is the pure end in itself, the pure joy of mere being.

But as we will see, this very identity contains enormous creative potential. The main activity of the pure end in itself is new - creation. The universal consciousness wants to recognize this global identity in detail, wants to know itself. To do this, it draws a first line. Drawing the boundary defines a subset, thereby (apparently) losing unity. One part (initially) faces another part. Any small subset of 'All That Is' (SMS 1) enters into a connection of a different kind than space-like with another subset (SMS 2). We call this different connection time-like (Figure 3.4b). A time-like communication emerges that initially appears cut off from space-like communication. Structure or space flickers (big bang, vacuum fluctuation) and the eternal here-and-now has another manifestation.

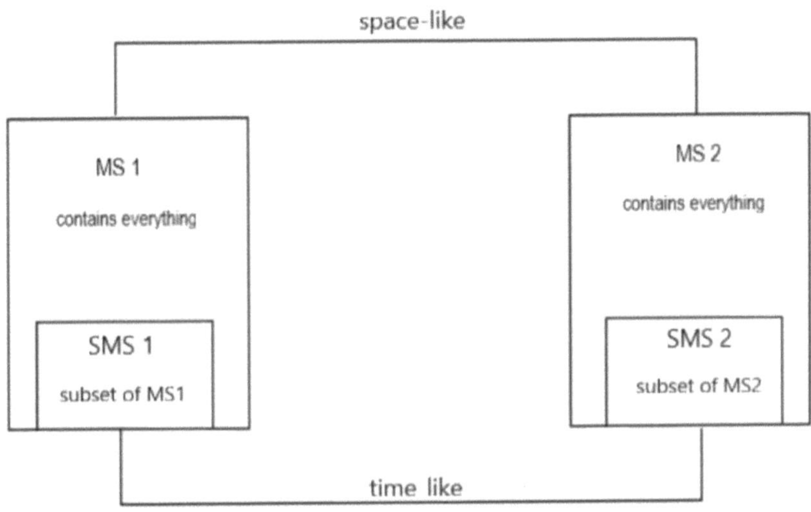

Fig.3.4b: The eternal here-and-now, 2nd phase

The subspaces strive to prove their identity to one another. You build optimal communication that ultimately confirms your identity. The subspaces develop into the macroscopic world we know and which produces humans. In the third phase, space-like communication also becomes conscious in the subspaces. Man discovers that the space-like connection has always existed. This characterizes the act of creation; a new unity has emerged, which is again identical to the all-unity and thus passes into the 4th phase, which is the same as the 1st phase.

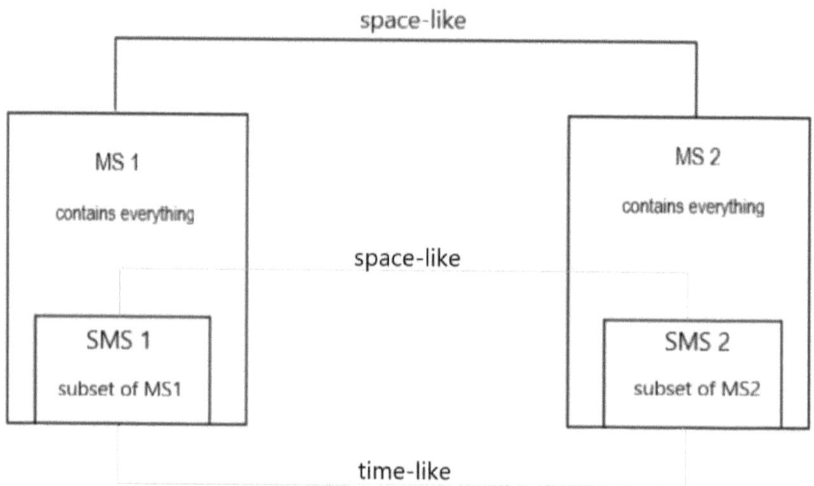

Fig.3.4c: The eternal here-and-now, 3rd phase

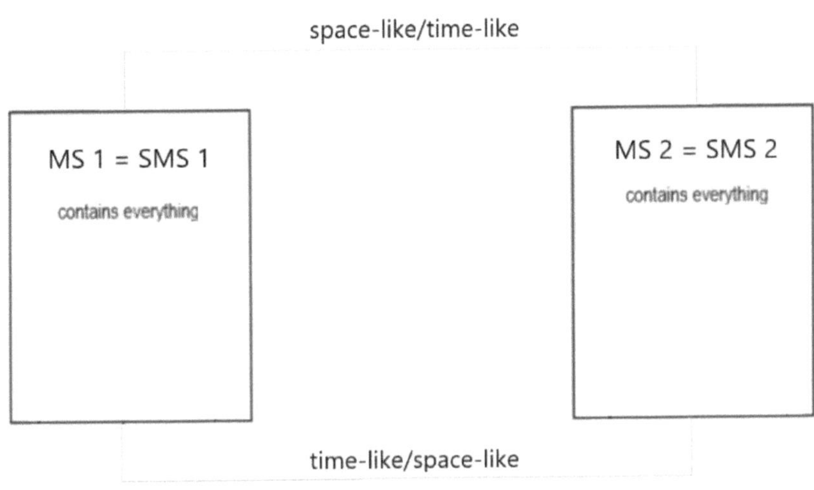

Fig.3.4d: The eternal here-and-now, 4th phase = 1st phase

MS 1 and MS 2 are constantly expanding as a result. Since the spatial connection between MS 1 and MS 2 always exists, MS 1 is always equal to MS 2.

The three phases shown cannot be seen one after the other, but rather all at the same time:
From the time-like perspective, unity is not immediately recognized because space-like communication is initially unknown. The time-like communicating units do not realize that their communication would not be possible without a space-like

connection. The nature of the first time-like interactions does not happen arbitrarily or randomly, but is coordinated by the still existing space-like connection between MS1 and MS2 (2nd phase). As a result, we are subject to the eternal laws of development of the ever-expanding universal consciousness.

Nevertheless, the impression temporarily arises as if there were two worlds, one and the other (subject/object). Violent communication leads to the conclusion that one part is identical to the other part. This means that temporal communication has become superfluous. In the newly created unity, the need to recognize oneself awakens again. By drawing further boundaries, it confronts itself again and the process begins again. Something new is constantly being created, but at the same time it is identical to the origin. This is only possible because of space-like communication.

We have already used the term *God* several times without - having established an exact context of meaning. At this point we could define the term *God*: *Omnipresence or universal consciousness is God.*

The omnipresence of God means that the new always remains identical to the old. Although the whole (God) is constantly changing, it remains in perfect identity with itself and its parts.

This definition of the term 'God' is the only one that makes sense to me.

Admittedly, we are talking about a model that is, in a sense, a closed system. There may well exist something outside this system that is unrecognizable to us. It could be something we don't even have a space-like connection to. Or it could be a connection that we can only recognize after we have overcome or left this world. The present model could then perhaps be expanded in the same way. There may be readers who would like to have the concept of God expanded to these levels. For my part, I believe that in this case it would be better not to define the concept of God at all. Because a definition only makes sense within a given context of meaning. As the Taoists said long time ago: You cannot give IT a name.

Again, in other words: Consciousness concentrates on a part of the whole and recognizes itself or the whole in this part.
The exclusion of a part from the whole initially creates the impression that this part is something other than the whole. This is exactly where time-like communication begins. The newly emerged local consciousness tries to find the whole in this part. We already know that he will succeed because the part contains the whole.
The local consciousness develops in such a way that this realization comes about. Local consciousness produces man. The moment at which man recognizes that the part contains the whole is the moment at which the part can be returned to the whole. The original boundary is transcended. Self-knowledge is complete. Creation is limiting and transcending. This process repeats itself constantly and everywhere. It is the constant will to self-knowledge.

The universe we know today, strictly following the principle of creation, produced humans, who in turn have a consciousness that is capable of self-knowledge. Complete self-knowledge is always the local end of the act of creation. Each local consciousness can independently end the act of creation. It has thereby returned to the whole and is available there for a new act of creation. Each local consciousness therefore has its own reality. Consciousness has recognized itself and is therefore able to take part in a new boundary-setting process. It is therefore in the renewed act of creation. Each local consciousness is the repository of its own history because it has known itself. The more consciousness transcends its limits, that is, expands, the more it is able to draw new boundaries, i.e. to be creative.

Personal consciousness

We remember that the great changes in physics began with the discovery of the duality of light. On the one hand, light was something concentrated, particle-like. Under other observation conditions it was also something infinitely extended and wave-like. We have also found this duality in consciousness, the time-like and the space-like channel. When we observe our personal consciousness, we notice an ability to concentrate on a local topic. We can completely identify with a thing; we can completely empathize with it; we focus. This is the time-like portion. But we can also relax our consciousness, expand it infinitely, so that we no longer think about anything and are not identified with anything. This brings us closer to the space-like part of our personal consciousness. In fact, consciousness, like light, has both parts at the same time. Only one part or the other always appears to us.

We now want to further expand the model of individual human consciousness established in Chapter 2 (Fig. 2.1). The time-like aspect of individual consciousness is the outward channel of knowledge. Free will controls this channel of knowledge and directs our attention to the desired perception. This process is called concentration. Relaxation, on the other hand, is when we direct the channel so that it is directed towards the spatial perception, i.e. inwards. In the limiting case it is not aligned with any perception.
Here we see that concentration and relaxation are similar concepts; In both cases what is meant is the direction of the channel of knowledge. The ability to concentrate and relax - increases with the accuracy of channel selection.

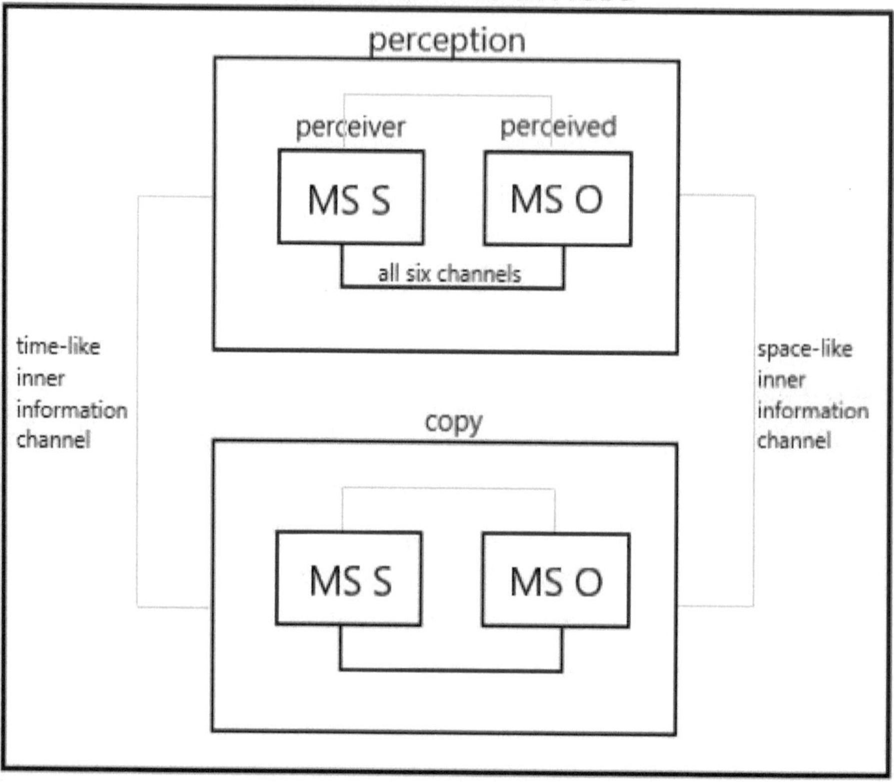

Figure 3.5:   *People recognize themselves*, love arises,
Cognitive ability and self-confidence are complete.

The more we are able to limit the time-like channel, the more the space-like channel becomes apparent. We already know that what really exists is influenced by what we imagine.

That is, our view of the world is primarily determined by the copy meaning space (Fig. 3.5). This in turn is determined by the ideas and beliefs that we have acquired.

So, let's ask ourselves what ideas and beliefs we have about our being. We imagine that we are a free individual integrated into an external world. This leads us to the separation between perceiver and perceived, between subject and object. Furthermore, we are convinced that we are endowed with free will, which controls the perceptions and actions of the individual. The will can help determine what we want to allow from the world into our incomplete meaning space copy, will thereby automatically and directly influences the nature of our reality.

We are used to free will controlling *what* we recognize: whether we turn to a tree or a person or simply our own thoughts by closing our eyes. What we are not used to, however, is that free will decides, *how* we perceive. In fact, there are many ways to recognize a tree, for example. From analytical dissection (biologist) to wood processing to a loving embrace, all types of communication with the tree are conceivable here. In children it can be observed that this *way of knowing becomes more and more restricted* during socialization. But this restriction is also a cosmic law. We specialize in a certain way of knowing in order to clearly distinguish ourselves from the unknowable and to produce the human form. In this way we can move in the unknown as a typically human form and thereby expand our consciousness.

The '*how*' of knowing corresponds to our own identity. Of course, there is a range of different ways of perceiving that are possible without us having to give up our own identity and the idea of the autonomous individual personality. The more often we change the '*how*', the more often we have to change and expand our identity.

A good way to carefully increase the range of '*how*' *is* to reduce the number of '*what*'. All possible perceptual activities are gradually shut down until, at the optimum, there is no longer any perception. Then the point in time has been reached at which the spatial influences can have their maximum impact. These then lead us with somnambulistic security to a new, stable identity. We perceive the world through different eyes. Our consciousness has expanded, we have really recognized something new. The quantity of knowledge is reduced in favor of a quality of knowledge. At this point we are led to love again. Love happens when a new quality of knowledge is achieved. A new quality of knowledge demands a change in identity. We need to expand our ideas about ourselves, this is what makes things so difficult for us, but also so exciting.
Love has both time-like and space-like aspects. An important prerequisite for our ability to love is therefore to be able to alternate concentration and relaxation and to be able to increase them to the respective depth. The depth, in turn, indicates how much we are able to precisely address and activate the different channels.

Our will is normally part of our self-consciousness and determines the respective communication channel. An interesting effect occurs during hypnosis. In this case, the will can be given to another person, i.e. to someone other than self-consciousness. This is an extreme case.

In the same way we can imagine that the will is distributed across different selves but still controlled by consciousness. With this idea we go beyond our model of self-consciousness and come to an idea of collective consciousness. This stage of development could be viewed as lying between self-consciousness and universal divine consciousness.

In fact, by distributing the will among a group of people, things can be achieved that we are unable to do as individuals. These are, for example, things that were previously described as miracles or magic. As long as not every member of the group is aware that there is a distributed will, those involved are also not aware that they themselves have a say in the events and therefore consider reality to be miraculous or at least incomprehensible. This effect can also sometimes be observed in team work.

Personal belief

Our believes are like the receiving frequency of a cosmic radio. I can receive or perceive exactly on the frequency that I set. Our personal beliefs are largely identical to our own meaning space, i.e. to our own ideas and beliefs. Belief shapes the space of meaning in such a way that certain information is received and others are not. According to our communication model, the information received must correspond to belief. If the information does not correspond to the belief/meaning space, we cannot receive or interpret it. Our personal perception is determined by our beliefs.

Our faith in the Universe is of existential significance. We tend to see the Universe outside of ourselves, assigning attributes to it, seeing ourselves as unsignificant, misbehaving and imperfect. We perceive the Universe and ourselves as something different. This perception determines our believes. This believes are characteristic for the Western view.

If we believe in God at all, it seems sufficient to us to believe in the existence of God. God exists independently of us. We are imperfect and must develop toward God. This is an unfavorable belief that brings us suffering. As Eastern religions and the consequences of our communication model show us, it is crucial to believe in the identity of God and ourselves. Belief is the state of our meaning space and this in turn determines our perception. Provided we have the right faith, we can perceive ourselves as identical with God. In the following chapter we will therefore try to get in touch with ourselves.

4   Inner perception: the path to the new inner world or Dialogue with God

We have expanded our communication model to a model of consciousness and even developed a theory of consciousness; we know the origin of all meanings. We can now open up all meanings spontaneously via the inner space-like channel and therefore always have an identical space of meaning available. We are filled with love, even then and especially when foreign information reaches us via the time-like channel. This has always been the case, but now we are only beginning to become aware of it, we are recognizing ourselves.

So, what happens when the *lover* wants to know himself? Our logical analytical mind tells us that this is not possible because the lover has already realized the identity with his counterpart. He has already recognized himself; he already knows everything about himself.

But what is happening in every moment is something different: He is constantly trying to find new boundaries. Behind these boundaries he suspects something other than himself. The mere assumption of this boundary is enough to make him fall into hectic activity. The other person needs to be examined in great detail in order to ultimately find out that he or she is identical to him or her.

This eternal process takes place in every moment. We are the divine aloneness. At the same time, we chase unity in order to discover that we are identical with it. Our chasing after ourselves is what creates time. From the first demarcation of the other to the determination that the other is identical with everything, including oneself, is what we call time passes. Anyone who frees themselves from time has found unity.

Now, of course, the question immediately arises as to why we isolate us in order to then establish an identity again. The answer is simply because it's fun. The fun of this world is proving unity over and over again. Once consciousness has found unity, it can be content with this state (the illuminated state) and thus end the fun. This happens all the time, but we are so in love with the fun and joy of being that we allow ourselves to be overtaken by doubts and isolation and therefore seek unity again. But actually, we never lost unity.

As strange as it may sound, the mere joy of being gives us temporality, finitude and suffering. Returning to unity is a purely conscious process (see below) and therefore possible at any time, anyone can stop the fun at any time. Leaving this world, i.e. the physical death, is *not* the return to unity. Death is only a transformation, a step towards unity, but not the achievement of unity. Achieving unity is much easier, it has always been achieved.

The fact that the fun can stop at any time, i.e. is potentially finite and therefore finding unity is possible *at any time*, removes any horror from temporality, finiteness and suffering. By taking away the horrors from our "worldly sufferings," they practically no longer exist. A great cheerfulness spreads that doesn't want to go away. So, we find that there really is a finite amount of cheerfulness. Finite joy is un-doubtfully better than endless suffering.

The consciousness process: After the identity of the other with everything has been established with cheerfulness (this can take varying amounts of time), the cheerfulness must give way. Timelessness occurs.

Meditation is a step towards timelessness, towards magic, unity. The step towards spatiality, dissolves everything. Everything dissolves into everything. This is extremely threatening, especially for a local consciousness. The local consciousness wants to hold on to its locality. And that is supposed to be fun, so holding on to it isn't a problem at all. On the contrary, locality is what makes the process of transition into spatiality possible. If the local consciousness were to completely accept its complete dissolution into everything, the gigantic divine creation process would be ended for this consciousness. All boundaries would be removed, a perfect identity would be established, unity would be found. The human form would be done with. How boring.
Fortunately, through meditation, for example, it is possible to briefly immerse yourself in unity, to see that unity has always been there and that life is therefore great fun. But let's now look at the meditation technique in detail.

The inner perception

Meditation in its fullest sense can perhaps be described as inner and outer perception at the same time or as an unconscious state (Figure 3.3). But this divine border case, which represents the perfect, all-encompassing love, is not meant here by meditation. Rather, what is meant is a technique of inner perception that can lead those for whom external perception is always in the foreground to perfection.

There are many schools and organizations that teach meditation techniques. I would like to briefly mention an organization that is particularly interesting from a scientific perspective; it is known as "Silva Mind Control".

This organization is represented all over the world and has specialized in the training of inner perception. The scientific approach here is that inner perception can be verified. Although it is a subjective verifiability, this subjective verifiability can also become objective at some point. (At the latest when everyone has had this subjective experience.)

This organization has invented many subjective perception techniques. I would like to briefly describe one of these subjective experiences, which can basically be verified by everyone:

A person goes into a state of relaxation (see Chapter 3, Communication with Nothingness). A second person gives the name, age and location of a third person who is not present and unknown to the relaxed first person.

The image of this third person spontaneously appears in the mind of the relaxed person. The person in a state of relaxation can describe the third person who is not present in detail and name their character traits and illnesses.

This technique of subjective perception impressed me the most because it is easy to test and repeat. *However, I must point out that the presentation given here is only suitable for verification after some preliminary exercises. These preliminary exercises are outlined below. However, a safe and successful check is probably only possible after further exercises under supervision.*

I will now give you a text that you can have read aloud slowly or speak on tape. It serves to transfer you into a relaxed state of consciousness, where inner perception works particularly well.

"Lie down or sit down completely relaxed. Close your eyes and first listen to your body. Are there any muscles that are still tense? Try to consciously tense these muscles and then relax them.
(20 sec break)
Now listen into your head. Are there any tensions to be felt? Are your eyes shaking? Try to completely let go of the facial muscles. Let the eyelids become very heavy until they are completely relaxed and closed. Your body is now completely relaxed.
(20 sec break)
Now listen into your thoughts. Put yourself in the shoes of a listener. Listen to your thoughts. Don't let any thought tempt

you to follow it. Remain a completely neutral observer. A wonderful feeling of lightness takes hold of you. You experience yourself as completely independent of everything.

What usually worries or stresses you is now far away. You see it from far away. The longer you stay in the position of observer, the weaker the thoughts become.

(two-minute break)

Your thoughts have calmed down. You are now in a state in which you can use your inner space-like perception.

In this state you have now reached, you can have the texts in the following chapter read to you. These texts are intended to be read aloud in a relaxed state of consciousness. They serve your deep inner understanding of the cosmic laws that underlie all our changes. They lead to positive changes in your life practice, you will be more successful and happier."

Another technique that is also successfully used by Silva practitioners is programming objects. Certain mental images are mentally linked to certain objects. Water, for example, is a popular programming object (holy water!). The idea that a particular patient is completely healthy again after drinking the programmed water is mentally transferred to the water. The water stores the information, so to speak, "Patient x will be completely healthy again if he drinks this water". When drinking, the patient absorbs this information and is influenced by it.

This effect can be easily explained using our model of consciousness. Water has a space of meaning as a communication partner. Through intensive communication with the water (via space-like channels), the specific meaning space of the water is changed. Our mental image is transferred to it. The water takes on a certain meaning that is firmly attached to it (see Chapter 1). How long water can store this information has not yet been precisely investigated. However, it was found that diluting the water does not affect the transmission of information.

The simplest and oldest form of inner perception is prayer, communication with God. Prayer is an inner dialogue conducted in deep relaxation and devotion. A dialogue with God. When your own thoughts have completely calmed down, then the time has come for dialogue with God.

The collective form of prayer, for example in a mass, can take place if the individual people have already experienced the inner dialogue and have had contact with God or have a strong belief in God. Collective belief represents a powerful programming technique, probably the strongest of all. The thoughts expressed in collective prayer create reality.

Examination of traditional prayers shows that reality today is shaped by these prayers. The closeness to God and hence to ourselves has already been lost. Let us consider the "Our Father" as an example:

Our Father, who art in heaven
Blessed be your name
Your kingdom come to us
Your Will will happen
As in heaven so also on earth
Our daily bread Give us today
And forgive us all our sins
Just as we forgive our debtors
And lead us not into temptation
But deliver us from evil
For yours is the kingdom and the power
And the glory forever. Amen.

The first line already creates a clear distance from ourselves. God is in heaven while we are on earth. Reinforced again by the third line: Your kingdom come to us.

The separation between heaven and earth is the classic problem of today's major Western religions. We are not aware that heaven has to be here on earth, right now. Guilt also still appears, although Christ came to "take away the sins of the world", to free us from original sin, from duality. His goal was and is to heal us from the suffering of any duality of good/evil, matter/spirit, etc.

Temptation is the "evil" that comes from outside. We are to be kept away from this and freed from it. This is a false assumption because we can only go through evil, we will only be able to suffer it and leave it behind, thus integrating duality.

A new version of the "Our Father" is produced below. I am convinced that this would produce a better reality.

Our God
Our God, You are with Us
Hallowed is Your name
We are your kingdom
The Will of all of us be done
Like here and everywhere
Daily meditation is safe for us
We forgive ourselves for our own mistakes
As well as those who make mistakes
We go through all temptations together
And survive all evil
Because here and now is the kingdom and the power
And the glory forever. Amen.

5 Summary of chapters 1-4 for reading aloud in deep meditation

The following texts can be read in sections or all at once in deep relaxation. Maybe you like one or the other text better, you can then use it more often.

We exist. Our environment exists. Our environment is limitless. Our environment has no end. We have no ultimate cause. We enjoy a completely baseless existence. Our reason for existing is completely open. We are unconditionally free. We live in infinite freedom. We are identical with All That Is.

There is space- and time-like interaction. Everything is connected to everything else via space- like interaction. - Everything is contained in everything. There is only one indivisible whole.
Every object we choose in this reality has the possibility of interacting with every other object - including those that lie in our future.

Matter has a tendency to exist. It has a tendency towards certain characteristics. Our own views determine what tendential properties these are.

Reality is made up of something that takes on exactly the properties that we attribute to it. There is not one *objective* reality, but there are *many* realities.

Every exchange of information requires interaction or communication. Communication and information gain lead to awareness. Communication is consciousness. Gaining information is expanding consciousness.
Prayer is the highest form of communication. Praying is communicating with All That Is. Praying is communicating with yourself. Praying is communicating with God. You are communicating with God in this moment!
Love unites man with the world. Love eliminates the separation of man from nature. Love unites us with All That Is.

Consciousness is communication. Consciousness has a time-like and a space-like communication channel. It can concentrate and relax. It constantly does this independently and thereby creates new realities. Consciousness is constantly growing.
We become aware that we have a consciousness. This allows us to participate in the constant creation of reality. When we change our consciousness, we change ourselves and reality.

Love and Physics

Consciousness is the ability of Energy and Matter to self-reference.

The self-referencing of Energy and Matter is Life.

Life gives birth to Space, Time, Gravity and other phenomena that we observe. Life consequentially leads to self-reflection.

Information is the self-referencing, meta-physical part of Conscious-ness. We have no detailed knowledge of how the self-referencing exactly works.

Growth is a basic principle of nature; it happens by itself and is therefore value-free. There is no "more" or "less" developed natural system. There is only constant transformation. God is movement and rest in one. We are not only connected to the divine origin, but identical with it. We are God.

We can become aware of our consciousness.

We are not our bodies, we are not our thoughts, we are not our image of ourselves. We are our consciousness.

We can become aware of our selves.

The world is an open system. There are endless possibilities for influencing our actions. The world and our actions are not predictable. The future is unpredictable. The future only exists as a possibility.

The amount of what can be experienced is arbitrarily large. There is only the path, which always includes the destination.

The goal is to increase love and joy. The path is change. Love and joy are the direction of change.

Suffering is not a wrong path, but a desired aspect of reality. It reflects our resistance to change. When we open ourselves to change, suffering disappears on its own and gives way to pure joy.

Every moment of the present fundamentally changes the world. Every thought and every action of every single person determines immediate world events. The very personal influence on world events cannot be overestimated. If my actions and thoughts change, I change. If I change, everything in the world changes. We are personally in a position to determine what happens in the world. And every person does this all the time.

Reality is not the sum of all human actions and ideas. There is no ONE common reality. There are just as many realities as there are consciousnesses. Every consciousness has its own reality, which is determined exclusively by the consciousness itself. It is therefore independent of the reality of any other consciousness, but affects all other realities. An infinite number of realities exist side by side at the same time.

Every single person has their own reality. God is identical with each of these realities. It connects everything together.

Everything is already included in everything. We *are* the divine aloneness. At the same time, we chase unity in order to discover that we are identical with it. Our chasing after unity is what creates time. From the first demarcation of the Other to the determination that the Other is identical with everything, is what we call 'time passed'. Anyone who frees themselves from time has found unity.

The fun in this world lies in discovering and proving unity again and again. The mere joy of being gives us temporality, finitude and suffering. Returning to unity is a pure consciousness process that is possible at any time, in which the fun, and thus also the suffering, naturally ceases. This possibility, which is always available, takes the horror out of suffering. A great sense of cheerfulness spreads.

Appendix

Literature recommendations

For the interested reader, I would like to point out three books
that can serve to further deepen the ideas at hand.
1. Fritjof Capra: "Turning Time"
2. Thorwald Dethlefsen: "Fate as an opportunity", Goldmann
3. Ken Wilber: "Paths to the Self"
The three books complement each other excellently because
the authors come from the various disciplines of physics,
psychology and philosophy.